Harvesting Fungi for beginners

A Comprehensive Guide to Oyster Mushroom Farming

Alex Morne

copyright@2024 Alex Morne all right reserved, no part of this book should be reproduced in any form or means without a prior written notice from the copyright holder

Contents

- Chapter One 4
  - Oyster Mushroom Farming 4
- Chapter Two 21
  - Farming outdoors versus indoors 21
- Chapter Three 29
  - Tools required 29
- Chapter Four 53
  - Harvesting methods 53
- Chapter Five 58
  - New developments in oyster mushroom farming technologies 58
- Chapter Six 74
  - Management of diseases and pests 74

Chapter One

Oyster Mushroom Farming

Brief details on oyster mushrooms

Pleurotus ostreatus, the scientific name for oyster mushrooms, is one of the most widely grown types of mushrooms in the world. They are highly valued for their nutritional value, mild flavor, and adaptability in the kitchen.

The characteristic oyster-shaped caps of oyster mushrooms are what set them apart; depending on the type, these caps might be white, gray, brown, or pink in color. Oyster mushrooms have short, off-center stems that are covered in gills that are usually white or cream in color.

In addition to being commercially grown, these mushrooms can also be found growing in the wild, frequently on hardwood trees including poplar, oak, and beech. On the other hand, they are also well-known for their capacity to grow on a range of substrates, including agricultural waste, sawdust, and straw.

In addition to their culinary merits, oyster mushrooms are valued for their nutritional content. They are abundant in protein, fiber, vitamins (especially B vitamins), minerals like potassium and iron, and low in calories and fat. Furthermore, bioactive substances like beta-glucans, which are thought to offer a number of health advantages including enhancing the immune system, are found in oyster mushrooms.

Growing in popularity among both home gardeners and commercial producers are oyster mushrooms because of their easy cultivation, quick growth rate, and large harvests. They can be substituted for meat in vegetarian and vegan cuisine, as well as utilized in stir-fries, soups, and pasta recipes. All things considered, oyster mushrooms are a tasty and healthy addition to the culinary scene, and they also bring fascinating prospects for local food production and sustainable agriculture.

Types of Substrates suitable for oyster Mushroom

One of the most adaptable mushroom species for cultivation is the oyster mushroom, which can grow on a variety of surfaces. The following are some typical substrates that work well for growing oyster mushrooms:

1. Straw: Oyster mushrooms are often grown on substrates such as wheat, rice, and other cereal straws. Straw gives mushroom mycelium a nice mix of nutrients and structure to invade.

2. Sawdust: Another common oyster mushroom substrate is hardwood sawdust, such as that from poplar, oak, or beech trees. For mushrooms to develop properly, sawdust needs to be supplemented with nitrogen-rich additives such as soybean meal or wheat bran.

3. Cottonseed hulls: Oyster mushroom growing can be successfully conducted on cottonseed hulls due to their high lignin content. To enhance texture and water retention, they are frequently used in conjunction with other substrates.

4. Corn cobs: Chopped or ground corn cobs can be used as an oyster mushroom substrate. They offer an excellent framework for mycelial growth and are reasonably simple to get.

5. Coffee grounds: Recycled coffee grounds are an easily accessible substrate for oyster mushroom farming. To improve texture and nutrient content, they should be supplemented with other ingredients like sawdust or straw.

6. Paper products: Shredded paper, cardboard, and paper pulp can all be utilized as oyster mushroom substrates. For mushroom cultivation to be successful, these resources need to be adequately pasteurized and supplemented with nitrogen sources.

7. Wood chips: Oyster mushrooms can be grown on hardwood chips made of beech or oak. Wood chips, on the other hand, need more time to colonize and might benefit from additions high in nitrogen.

8. Agricultural by-products: Oyster mushrooms may grow on a variety of agricultural waste products, including corn stover, sugarcane bagasse, and cotton waste. Pre-processing and supplementation may be necessary for these materials in order to improve nutritional availability.

In order to avoid contamination and guarantee good mushroom growing, surfaces must be well cleaned and prepared before being inoculated with oyster mushroom spawn. Additionally, gardeners may experiment with different substrates to find the best outcomes for their particular conditions, as this can have an impact on the productivity, quality, and flavor of the harvested mushrooms.

Requirement for Infrastructure

The infrastructure needed for oyster mushroom farming varies depending on the size of the business, the type of cultivation method (indoor or outdoor), and the surrounding circumstances. An outline of the facilities required for oyster mushroom farming is provided below:

1. Expanding Area:

• **Indoors:** You'll need a specific indoor area, like a room, cellar, greenhouse, or warehouse. The area needs to have enough natural or artificial lighting, ventilation, and temperature management.

• **Outside:** A shady space shielded from harsh weather and direct sunshine is required for outdoor growing. The necessary shade can be produced with shade cloth or by the natural shading provided by trees.

2. Growing Beds or Containers:

- **Containers:** You can cultivate oyster mushrooms indoors using grow bags, trays, buckets, or plastic bags. These should be sturdy, hygienic, and big enough to hold the substrate.

- **Beds:** Raised beds or ditches filled with substrate can be used for outdoor gardening. In order to avoid water logging, beds should be prepared with appropriate drainage.

3. Equipment for Preparing Substrates:

- **Chipper or grinder:** Used to reduce raw materials into smaller pieces, like wood chips or straw.

- **Mixing tools:** To fully blend substrates and supplements, use a tumbler or mixer.

- **Pasteurization apparatus:** steam generators or large tanks used to heat and pasteurize substrates in order to destroy competing bacteria.

4. Sterilization Tools:

- **Autoclave or pressure cooker:** To avoid bacterial, fungal, and other microorganism contamination, sterilize containers, substrate, and instruments.

- **Chemical sterilizers:** Bleach solutions or hydrogen peroxide can also be used to disinfect surfaces.

5. Areas of Incubation and Inoculation:

- **Laminar flow hoods or clean rooms**: Used to inoculate substrate with mushroom spawn in a sterile setting to avoid contamination.

- **Incubation room or chamber**: To keep the ideal humidity and temperature for mycelial colonization of the substrate following inoculation.

6. Systems for controlling the climate:

- **HVAC systems:** These are used, particularly in indoor growth setups, to control the humidity and temperature within the growing environment.

- **Humidifiers or fogging systems**: To sustain the high humidity conditions necessary for the growth of oyster mushrooms, especially when they are fruiting.

7. Area for Harvesting and Processing:

- **Workbench or table:** Used for gathering mushrooms and getting them ready for canning or preserving.

- **Equipment for cleaning and packing:** Scales, brushes, knives, and packaging materials are used to clean, sort, and package gathered mushrooms.

8. Facilities for Storage:

- **Cold storage, or refrigeration:** To prolong the shelf life of harvested mushrooms.

- **Dry storage:** Keeping farming supplies, supplements, and substrates in a dry, well-ventilated space.

9. Services and Utilities:

- **Water availability:** For cleaning tools, preserving

humidity levels, and moistening substrates.

- **Electricity:** Used to run air conditioning and heating units, lights, sterilise objects, and other electrical appliances.

- **Waste management:** The appropriate removal of used substrates and other leftovers from the cultivation process.

10. Supplies and Equipment for Safety:

- **Personal protective equipment (PPE)**: To guard against chemical exposure and biohazards, wear gloves, masks, and lab coats.

- **First aid kit:** To treat small wounds and mishaps that could happen while farming.

- **Fire extinguishers:** To address electrical and heating equipment-related fire dangers.

Chapter Two

Farming outdoors versus indoors

The two main techniques for growing oyster mushrooms are indoor and outdoor farming, each with unique benefits and drawbacks. Here's a comparison of oyster mushroom farming conducted outside and indoors:

Indoor Agriculture:

1. Controlled Environment: Growers can precisely regulate environmental parameters including light, humidity, and temperature when they cultivate indoors. Throughout the year, improved quality and more consistent yields of mushrooms can result from this control.

2. Year-Round Production: Growers may produce oyster mushrooms all year round, independent of the outside weather, by using indoor growing. This guarantees a consistent harvest of mushrooms and can be especially helpful in areas with harsh weather.

3. Pest and Disease Protection: Indoor spaces offer some defense against diseases, pests, and other environmental elements that could harm crops grown outdoors. This lowers the possibility of crop losses brought on by pest infestations or unfavorable weather conditions.

4. Space Efficiency: Growers can increase production in a small space by using indoor farming, which can be space-efficient. Shelving units, stacked containers, and vertical farming techniques can all be used to maximize space use and boost production capacity.

5. Greater Initial Investment: The infrastructure needed to establish an indoor mushroom farm includes lighting, sterilizing equipment, and climate control systems. Over time, nevertheless, these investments may result in more consistent production outcomes and higher yields.

6. Labor-intensive: Activities including substrate preparation, inoculation, and environmental condition maintenance may call for additional labor while cultivating mushrooms inside. Mechanization and automation, however, can assist save labor expenses and improve operational efficiency.

Farming outdoor:

1. Lower Initial expenditure: Compared to indoor farming, outdoor farming sometimes requires smaller initial infrastructure expenditure. To cut expenditures up front, modest constructions like raised beds or shade structures could work just well.

2. Natural Light Source: Sunlight is the most abundant natural light source for outdoor growth. This may lead to a more sustainable operation and lower energy expenses related to artificial lighting used in indoor farming.

3. Possibility of Large-Scale Production: Growing outdoors, particularly in open fields or designated growing spaces, presents the possibility of large-scale production. Commercial farmers looking to expand their operations or fulfill strong demand may find this scalability useful.

4. Seasonal Production: Depending on the weather, outdoor farming is frequently seasonal. Production can be restricted to particular seasons of the year, which could lead to variations in supply and possibly reduced yields in unfavorable weather.

5. Vulnerability to Environmental Factors: Crops grown outside are more susceptible to pests, wind, rain, and temperature swings. Growers might have to employ tactics to lessen these hazards, like employing integrated pest management techniques or protective covers.

6. Possibility of Organic Certification: Since producers can use natural inputs like compost and organic substrates, outdoor production might be more suited to organic farming techniques. Customers looking for organic or responsibly farmed mushrooms might find this appealing.

The decision to cultivate oyster mushrooms indoors or outdoors ultimately comes down to a number of variables; including grower preferences, market demand, climate, and resource availability. While some producers may favor the lower initial investment and scalability of outdoor farming, others may choose indoor farming for its precise control and year-round production. Furthermore, some farmers can combine the two approaches to diversify their output and lower risks.

Chapter Three

Tools required

1. Sterilization Tools:

● Autoclave or pressure cooker: To avoid contamination, sterilize substrates, receptacles, and instruments.

● Chemical sterilizers: For surface sterilization, use solutions of bleach or hydrogen peroxide.

2. Equipment for Preparing Substrates:

● Chipper or grinder: Used to reduce raw materials into smaller pieces, like wood chips or straw.

- Mixing tools: To fully blend substrates and supplements, use a tumbler or mixer.

- Pasteurization apparatus: steam generators or large tanks used to heat and pasteurize substrates in order to destroy competing bacteria.

3. Instruments for Inoculation:

- Scalpel or knife: For inoculating substrate and chopping mushroom spawn.

- Syringe or inoculation loop: For injecting mushroom spawn into the substrate.

- Laminar flow hoods or clean rooms: Used to inoculate substrate in a sterile setting to avoid contamination.

4. Equipment for Growing and Incubation:

- Incubation chamber or room: To keep the ideal humidity and temperature for the substrate's mycelial colonization.

- Shelving systems or racks: Used to arrange and arrange trays or containers of infected substrate.

- Heating mats or pads: Particularly in colder climates, to

offer warmth throughout the incubation phase.

- Humidifiers or fogging systems: To sustain the high humidity levels necessary for the growth of mushrooms, especially when they are fruiting.

- Fans or ventilation systems: To provide appropriate airflow and avoid carbon dioxide accumulation.

5. Equipment for Harvesting and Processing:

- Spotlessly clean workbench or table: For gathering mushrooms and getting them ready for canning or chopping.

- Soft towels or brushes: To clean and remove debris from gathered mushrooms.

- Packaging supplies: For example, trays, bags, or containers to package mushroom.

- Scale for weighing: Used to portion and measure gathered mushrooms.

6. Systems for controlling the climate:

- HVAC systems: These are used, particularly in indoor growth setups, to control the humidity and temperature within the growing environment.

- Sensors for temperature and humidity: To keep an eye on the surroundings and adapt as necessary.

7. Facilities for Storage:

- Cold storage, or refrigeration: To prolong the shelf life of harvested mushrooms.

- Dry storage: Keeping farming supplies, supplements, and substrates in a dry, well-ventilated space.

8. Supplies and Equipment for Safety:

- Personal protective equipment (PPE): To guard against chemical exposure and biohazards, wear gloves, masks, and lab coats.

- First aid kit: To treat small wounds and mishaps that could happen while farming.

- Fire extinguishers: To address electrical and heating equipment-related fire dangers.

9. Services and Utilities:

- Water availability: For cleaning tools, preserving humidity levels, and moistening substrates.

- Electricity: Used to run air conditioning and heating units, lights, sterilize- objects, and other electrical appliances.

- Waste management: The appropriate removal of used substrates and other leftovers from the cultivation process.

10. Tools for Tracking and Documentation:

- pH meter: Used to track the pH of supplements and substrates.

- Hygrometer and thermometer: Used to gauge humidity and temperature.

- Software or a logbook: Used to keep track of important information including harvest yields, growth milestones, and vaccination dates.

Creating the growing area

When designing a growing facility for oyster mushroom production, the goal is to maximize space usage and operating efficiency while fostering ideal development circumstances. Here is a detailed guide to help you plan the growing area:

1. Select the Right Space:

• Pick a spotless, well-ventilated location with sufficient access to utilities like electricity and water.

• Take into account elements like managing temperature, having access to natural light, and being close to other farming activities.

2. Comparing Indoor and Outdoor Cultivation:

The choice of growing mushrooms indoors or outdoors depends on a number of variables, including temperature, space, and resource availability.

• Although it costs more to set up the infrastructure, indoor cultivation offers more control over the atmosphere.

• Although it could be less expensive, outdoor production is more vulnerable to environmental and meteorological changes.

3. Arrangement Scheduling:

- Draw a plan of the growth area's layout, taking into account things like equipment location, workflow, and space use.

- Set aside areas for preparing the substrate, inoculating, incubating, fruiting, gathering, and packaging.

4. Infrastructure Needs:

- Set up tables, racks, or storage to accommodate growing beds or containers. Think about making the most of vertical space to increase output.

- Make sure there is enough room between shelves or racks to allow for airflow and easy access.

- To reduce the danger of contamination, set up a clean room or other designated space for sterile procedures like immunization.

- If indoor cultivation is selected, install lighting fixtures. The ideal lighting conditions for mushroom growth can be achieved with LED grow lights that have adjustable spectrum.

5. Systems for controlling the climate:

- Install HVAC systems (heating, ventilation, and air conditioning) to control humidity and temperature, particularly inside.

- To maintain high humidity levels during the fruiting stage, use fogging devices or humidifiers.

- Use sensors and automated climate control systems to monitor and manage environmental parameters.

6. Water Source and Drainage:

- Make sure water sources are easily accessible for operations like cleaning, watering, and substrate preparation.

- Install drainage systems to provide adequate water drainage and prevent water logging.

7. Equipment for Pasteurization and Sterilization:

• Prepare sterilizing apparatus, such as autoclaves or pressure cookers, to sanitize containers, tools, and substrates.

• Set aside a specific space for pasteurizing substrates in hot water baths or steam.

8. Aspects related to safety and hygiene:

• Follow the right hygienic procedures to reduce the chance of infection.

Enough ventilation and personal protection equipment (PPE)

should be made available to employees who handle chemicals or operate in enclosed locations.

• Place first aid supplies and fire extinguishers in easily accessible areas.

9. Both ergonomics and accessibility:

• Plan the arrangement to reduce needless mobility and maximize workflow.

• Make sure that chores involving maintenance and harvesting may conveniently access the growing areas and equipment.

• Take ergonomic considerations into account to reduce worker strain and tiredness during regular activities.

10. Scalability and Prospective Growth:

• Consider scalability while designing the layout to allow for future growth or adjustments in production volume.

• As the business expands, make space for more equipment, a larger growth area, or infrastructure improvements.

Growers can create a well-organized and productive growing place for oyster mushroom growth by following these procedures and taking important elements like climate control, workflow efficiency, and space use into consideration.

Sanitation and hygiene guidelines

In order to avoid contamination and guarantee the operation's success, oyster mushroom cultivation requires strict adherence to hygiene and sanitation regulations. This is a thorough guide to sanitation and hygiene practices:

1. Individual cleanliness:

- Make sure that everyone who works with substrates or handles mushrooms maintains proper personal hygiene, which includes washing their hands often with soap and water.

To reduce the possibility of introducing contaminants into the growing area, provide personnel with hygienic uniforms or protective clothes.

2. Sterile area or clean room:

- To reduce the chance of contamination, set up a clean room or sterile space for activities including spawning, inoculation, and substrate preparation.

- To aid with cleaning and hygienic conditions, keep the

clean room clear of superfluous objects and clutter.

3. Sanitizing and disinfecting:

• Work surfaces, containers, shelves, and utensils should all be cleaned and disinfected on a regular basis when it comes to the tools, equipment, and surfaces used in mushroom production.

• Use sanitizers or disinfectants that have been approved and are effective against a wide range of viruses, fungus, and bacteria.

• make a clean schedule and ensure all cleaning procedures are often performed.

4. Containers and Substrates Should Be Sterilized:

• Before using, sterilize instruments, containers, and substrates to get rid of any possible contamination sources.

• Sterilize substrates with autoclaves, pressure cookers, or other sterilizing apparatus; alternatively, pasteurize them in steam or hot water baths.

• Ensure that the substrates are kept clean during the inoculation process by adhering to recommended sterilizing protocols.

5. Ventilation and Air Filtration:

• Install air filtration devices, including HEPA filters, to reduce airborne pollutants in the space used for cultivation

• Make sure there is enough ventilation to keep the air moving properly and avoid stale air accumulation, which can contain pollutants.

6. Management of Pests and Diseases:

• Use integrated pest management (IPM) techniques to keep the growing area free of pest infestations.

- Keep a regular eye out for symptoms of illnesses and pests, and act quickly to eradicate infestations when necessary by applying chemical, cultural, or biological management measures.

7. Management of Waste:

- To avoid the accumulation of possible pollutants, dispose of used substrates, contaminated materials, and other waste products created during growing properly.

- Adhere to local trash management and disposal regulations and use designated garbage disposal receptacles or containers.

8. Employee Education:

• Offer thorough instruction on hygiene and sanitation practices, such as appropriate hand washing methods, cleaning methods, and contamination control measures, to every employee.

• Stress how crucial it is to follow hygienic and sanitation guidelines in order to preserve the integrity of the growing environment and guarantee the caliber of the mushrooms that are produced.

9. Frequent Inspections and Assessments:

• Regularly audit and examine the growing area to make sure

cleanliness and hygiene regulations are being followed.

• Keep an eye on the weather, including temperature, humidity, and air quality, and note any changes that would point to possible contamination hazards.

Chapter Four

Harvesting methods

Care must be taken when harvesting oyster mushrooms to guarantee the best quality and to encourage ongoing mushroom production.

1. Time and Observation:

- Consistently track the development of oyster mushrooms. When the caps have fully formed but before they begin to curl upwards or flatten at the margins, harvest them.

- Since oyster mushrooms grow quickly, once they start to fruit, keep an eye out for overly developed specimens by checking them every day.

2. Careful Elimination:

• Cut the mushrooms at the base of the stem using a sharp knife or pair of scissors. To prevent harming the mycelium and influencing subsequent flushes, refrain from tugging or twisting the mushrooms.

• To reduce waste and promote the emergence of new flushes, trim the mushrooms just below the surface of the substrate.

3. Gathering in Groups:

• Gather mushrooms as they get ripe in batches rather than

waiting for the entire crop to do so at once. This makes it possible to harvest continuously over a few days or weeks.

- Cut out the larger mushrooms, leaving the smaller ones to grow.

4. Clean the Tools:

- To reduce the chance of infection, sanitize harvesting instruments (such as knives or scissors) with alcohol or a diluted bleach solution before to harvesting.

- To stop infections from spreading, clean instruments after every harvesting session.

5. Preserve Hygiene:

• Before handling mushrooms, fully wash your hands to stop the spread of pollutants.

• To reduce the chance of contamination, wear clean gloves and refrain from touching the mushrooms directly with your bare hands.

6. Prevent Bruising:

• Take care while handling mushrooms to prevent bruising or harming the fragile caps, since this may compromise their look and shorten their shelf life.

• To avoid bruising and moisture buildup, place picked mushrooms in a shallow container or basket lined with clean paper towels.

7. After-Harvest Management:

Sort the obtained mushrooms according to their size, form, and quality. Take out any mushrooms that are broken or discolored.

Mushrooms should be kept cool, in a place with enough ventilation and air flow to keep them fresh and stop mold from growing.

• To prevent compression and deterioration, don't stack mushrooms on top of one another.

Chapter Five

New developments in oyster mushroom farming technologies

New technologies are always changing the oyster mushroom industry and providing creative ways to increase production, sustainability, and efficiency. Among the noteworthy new technologies being used in oyster mushroom growing are:

1. Automated Surveillance Systems:

• Environmental parameters like temperature, humidity, CO_2 levels, and substrate moisture content are tracked in real-time using sensor-based monitoring systems.

- Growers can maintain ideal growth conditions and make data-driven decisions for better crop management with the help of these technologies.

2. Accurate Climate Control:

- Sophisticated climate control systems use machine learning algorithms and data analytics to accurately control the ambient conditions in crop facilities.

- These systems maximize energy efficiency, reduce resource waste, and cultivate the perfect environment for oyster mushroom growth.

3. Aeroponic farming and vertical farming:

- Vertical farming techniques increase production density by stacking growing trays or containers vertically, which maximizes the use of available space.

- Research is being done on aeroponic systems, which cultivate mushrooms without soil or substrate, to further improve space efficiency and lower resource usage.

4. Advanced Formulations for Substrates:

The goal of the research is to create new substrate compositions with locally and sustainably derived resources.

- As an alternative to conventional substrates, biodegradable and renewable materials like coffee grounds, leftover brewery grains, and agricultural waste are being investigated.

5. Strain selection and genetic engineering:

- Oyster mushroom strains that are resistant to disease and have a high yield are being created through genetic engineering.

- To find desirable features and enhance the performance of commercial oyster mushroom cultivars, selective breeding procedures and genomic analysis are employed.

6. Technologies for Fermentation and Bioreactors:

- For the large-scale generation of mushroom spawn and mycelium, bioreactors with regulated agitation, aeration, and nutrient supplementation are employed.

- The breakdown of complex organic matter and improved nutrient availability for mushroom production are achieved by fermentation methods, which optimize substrate preparation.

Cultivators of oyster mushrooms can overcome current obstacles, increase production efficiency, and promote sustainable agricultural practices by adopting these cutting-edge technologies.

Fruiting stage supervision

In oyster mushroom production, controlling the fruiting stage is essential to achieving the best possible growth, development, and yield. This is a thorough guide to managing the fruiting stage:

1. Fruiting Has Started:

• Once the mycelium has colonized the substrate and the environmental circumstances are favorable for fruiting, fruiting initiation usually takes place. A number of variables, including light, humidity, temperature, and the exchange of fresh air, can cause this stage.

• Fruiting can be accelerated by lowering the temperature and raising the humidity. Opening vents or modifying airflow to provide a fresh air exchange encourages the induction of primordia, or pinning—the first stages of mushroom production.

2. Keeping the Environment at its Ideal:

• Temperature: During the fruiting period, keep the temperature between 18 and 24°C (64 and 75°F), with small variations to reflect the natural environment. Warmer temperatures increase mushroom development at a quick pace, whereas cooler temperatures encourage pinning.

• Humidity: To avoid drying out the mushrooms and encourage healthy growth, keep the relative humidity (RH) high, between 85 and 95 percent. Humidity levels can be maintained with frequent misting or humidifier use.

• Light: Although they are not as light-sensitive as certain other mushroom species, oyster mushrooms nevertheless need to be exposed to some light for healthy growth. One can use artificial light sources, such as low-intensity fluorescent or LED lights, or indirect natural light.

• Fresh Air Exchange: Make sure there is enough air movement to encourage the evacuation of the metabolic gases that the mushrooms release and to prevent the accumulation of carbon dioxide. To avoid problems like carbon dioxide toxicity or mushroom mold infection, proper ventilation is crucial.

3. Controlling Humidity:

Regularly monitor humidity levels and make necessary adjustments to avoid excessive drying or accumulation of moisture. Proper air circulation and high humidity levels help keep mushroom tops from getting soggy and encourage even development.

- To generate a microclimate around fruiting trays or containers, especially in dry environments, use humidity tents or plastic sheeting.

4. Watering and Controlling Wetness:

• Care should be taken during watering to prevent oversaturation of the substrate or wet circumstances, which might result in disease-causing mushrooms or low-quality mushrooms.

• Water mushrooms sparingly, concentrating on the surface of the substrate rather than the mushrooms themselves, using a fine mist or spray bottle.

5. Attachment and the Growth of Mushrooms:

● Keep an eye on the development of pins and modify the surrounding environment as necessary to encourage uniform pinset creation. Variations in humidity, airflow, or temperature can cause uneven pinning.

● To prevent competition for space and resources and to allow mushrooms to grow to their maximum potential, space out fruiting trays or pots adequately to prevent congestion.

● To encourage the growth of healthy, uniform mushrooms,

remove any aberrant or misshapen mushrooms.

6. Gathering:

• When mushrooms are at their ideal size and maturity, harvest them. When the caps are fully formed but before they begin to flatten or curl upward at the margins, oyster mushrooms are usually harvested.

• Slice mushrooms at the base of the stem using a sharp knife or pair of scissors. Refrain from tugging or twisting as this may harm the mycelium and interfere with subsequent flushes.

- As mushrooms ripen, harvest them in batches to maintain productivity and avoid over ripening.

7. Management after Harvest:

- Clear the growing area of spent mushroom substrate, or substrate that has finished fruiting, to avoid contamination and preserve hygienic conditions.

- Properly dispose of wasted substrate by composting or using other environmentally friendly techniques.

- To stop the spread of pollutants, clean and sterilize

harvesting instruments and equipment in between batches.

8. Observation and Documentation:

- Throughout the fruiting stage, keep a regular eye on environmental factors including temperature, humidity, and CO_2 levels.

- Maintain thorough records of the lengths of the fruiting cycles, harvest yields, and any observations or problems that arise. This data can be used to solve issues and optimize management of the fruiting stage in the future.

Growers may promote sustainable and ecologically friendly techniques in oyster mushroom farming while optimizing yield, quality, and profitability through efficient management of the fruiting stage.

Chapter Six

Management of diseases and pests

Controlling pests and diseases is essential to maintaining healthy crops and achieving maximum yields while growing oyster mushrooms. The following are some practical methods for controlling illnesses and pests in the production of oyster mushrooms:

1. Preventive actions:

- To reduce the chance of bringing pests and diseases into the growing environment, start with clean, premium spawn and substrate.

- Put in place stringent hygienic practices, such as routinely washing and disinfecting the cultivation area, growth containers, and equipment.

- When feasible, choose substrates that are less likely to become contaminated and use pest-resistant mushroom strains.

2. Integrated Veterinary Practices (IPM):

- Use an Integrated Pest Management (IPM) strategy to reduce the use of chemical pesticides by combining several pest management techniques.

- Keep a close eye on the developing environment for indications of diseases like mold or bacterial infections, as well as pests like insects, mites, and other arthropods.

- Establish cultural controls to prevent the growth of pests and diseases, such as maintaining appropriate humidity levels, air circulation, and cleanliness techniques.

3. Biological Regulators:

- Introduce microbial antagonists, parasitic organisms, or natural predators that are tailored to certain infections or pests.

Predatory mites, which manage mite infestations, and advantageous fungus such as Trichoderma spp. to prevent the growth of mold.

4. Chemical controls (last option):

• Adhere to label instructions and safety requirements when using chemical pesticides or fungicides, and use them sparingly and only when necessary.

• Select goods with low environmental effect and low toxicity to non-target creatures.

• Switch between various chemical classes to stop pest

populations from becoming resistant to pesticides.

5. Exclusion and Quarantine:

● Keep incoming supplies, like equipment and substrate, in quarantine to stop the spread of illnesses and pests from outside sources.

● Install bug screens on windows and entrances and restrict access to the cultivation area to authorized staff only.

Through the application of these pest and disease control techniques, oyster mushroom cultivators can limit crop losses, lessen their need on chemical inputs, and uphold ecologically sustainable production methods.

www.ingramcontent.com/pod-product-compliance
Lightning Source LLC
Chambersburg PA
CBHW070357230526
45471CB00006B/2618